THE ZAMBIT INEQUALITY DREAM

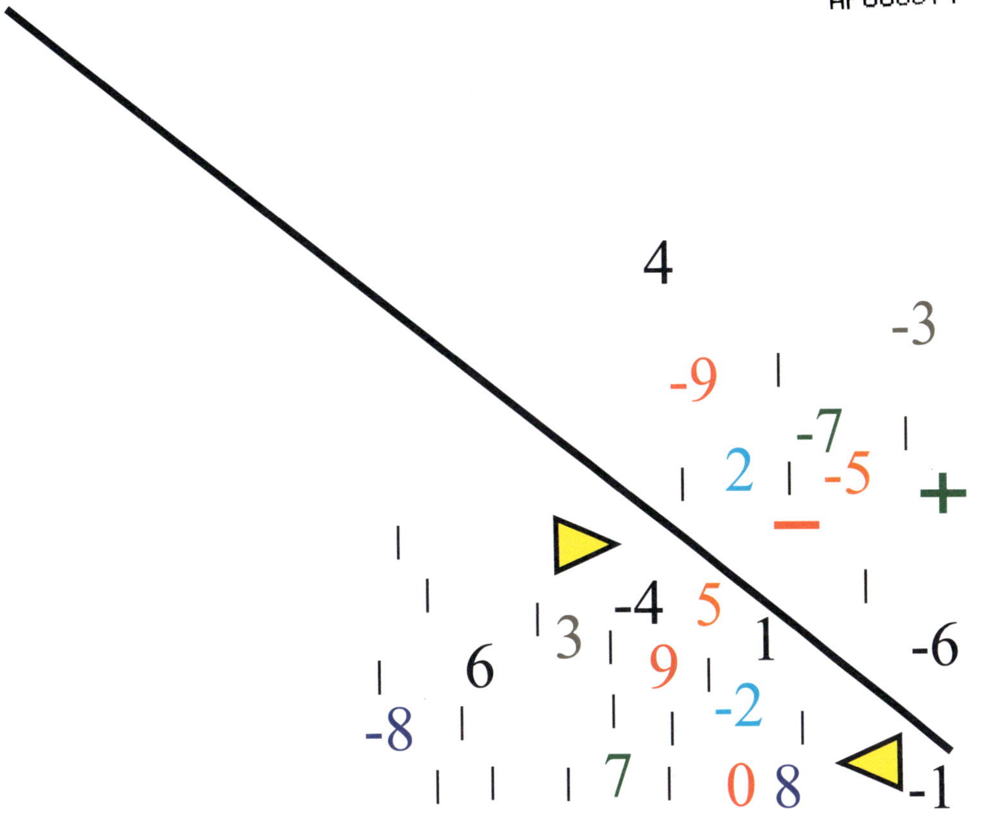

KlevaKids

Published by KlevaKids.com Inc

New York, New York

For more titles,visit our website at www.klevakids.com

KlevaKids

KlevaKids

Presents

THE ZAMBIT INEQUALITY DREAM

This story introduces the basic concepts of number line inequalities in an enjoyable format.

The above statements are purely informational. The purpose of the book is not to teach, but to expose concepts. Read the book to a child and just allow the story to be enjoyed.

There was a pile of stuff

just sitting around.

A way to organize it

had to be found.

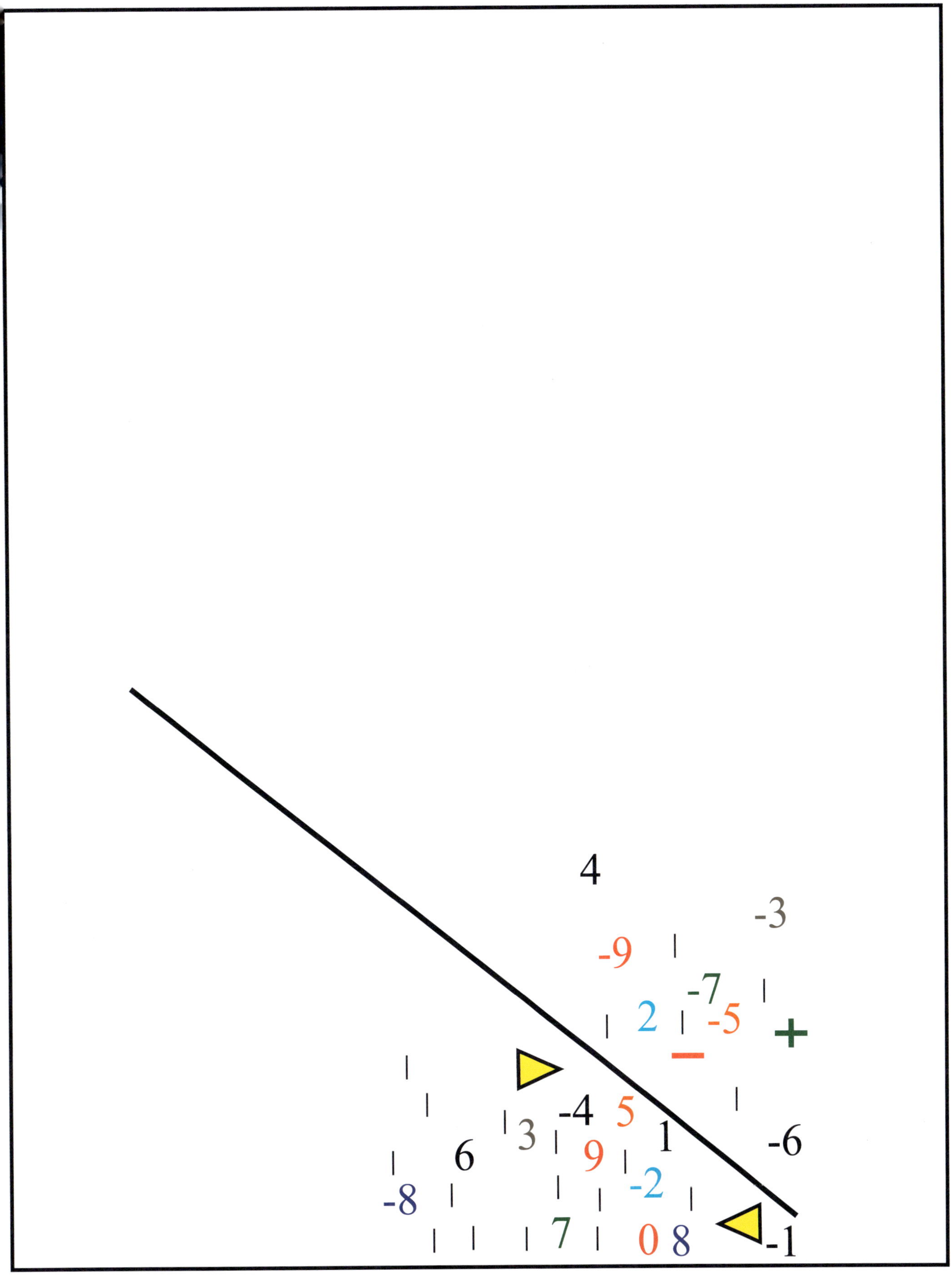
4
-3
-9
-7
2
-5
+
−
-4
5
3
1
-6
6
9
-2
-8
7
0
8
-1

The math symbol contraption arrived

and was put to the test.

Organizing piles of things

is what it does best.

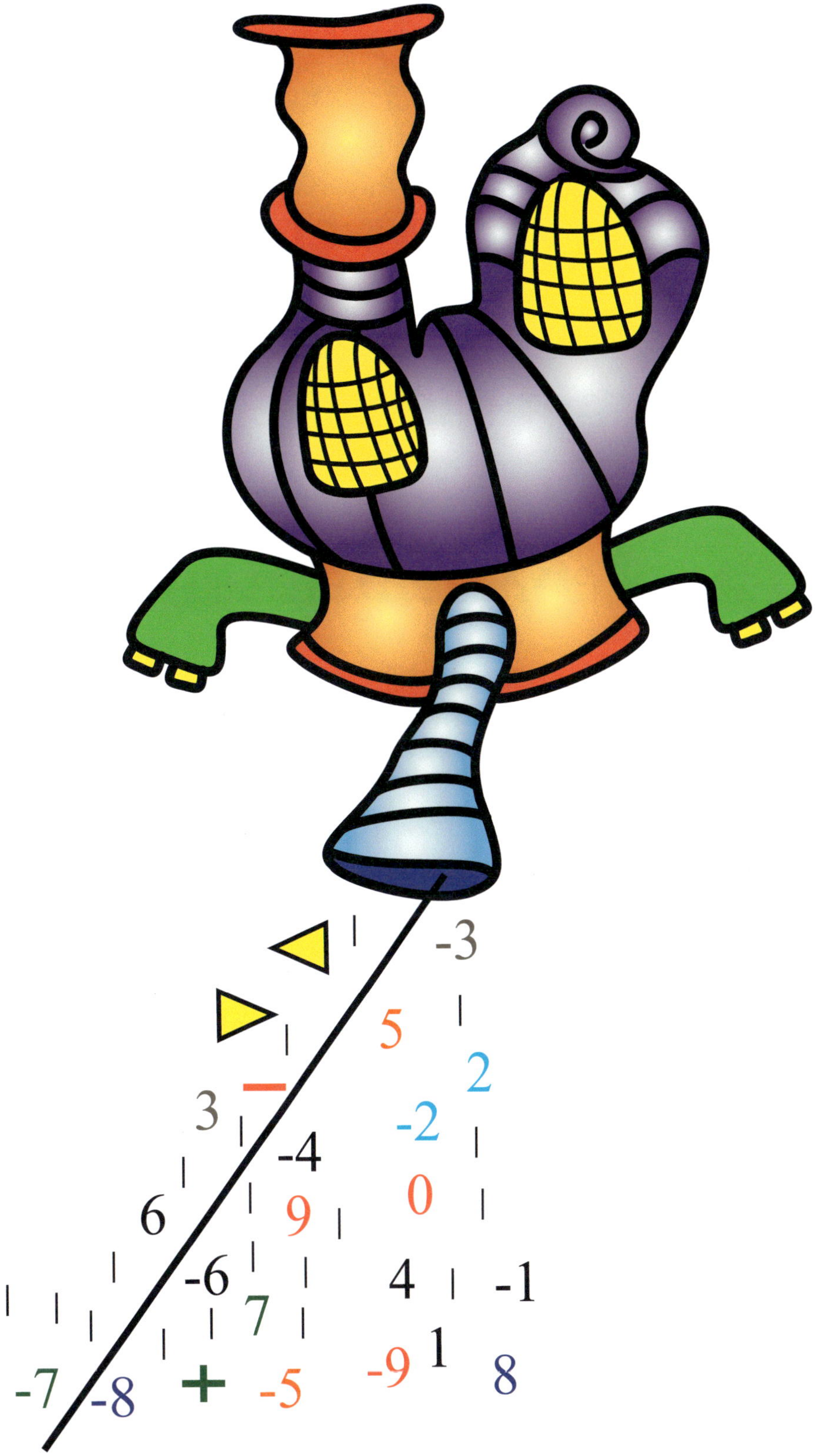
-3
5
2
3
−
-2
-4
0
6
9
-6
4
-1
7
1
-7
-8
+
-5
-9
8

Double arrows emerged

as part of the collection.

They represented infinity

in each direction.

The next things to come

were a bit of a trick.

They are called hash marks

or sometimes a tick.

Next came zero.

Where it went was no riddle.

It had to be at the origin,

exactly in the middle.

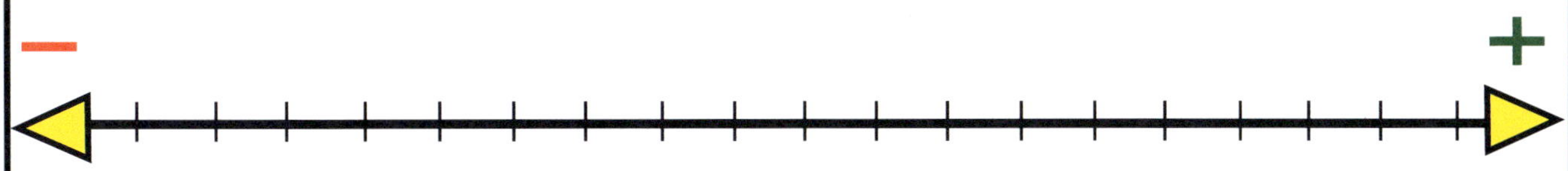

0

That delightful contraption

was so full of grace.

Positive numbers one through nine

also had a place.

–
+
0
1
2
6
3
7
4
8
5
9

Out came negative numbers

in an orderly form.

Placement to the left of zero

was the usual norm.

–
+
-1 0 1 2 3 4 5 6 7 8 9
-2
-3
-4
-7 -5
-8
-9 -6

The contraption went away,

but that was really fine.

In front of me

was a perfect number line.

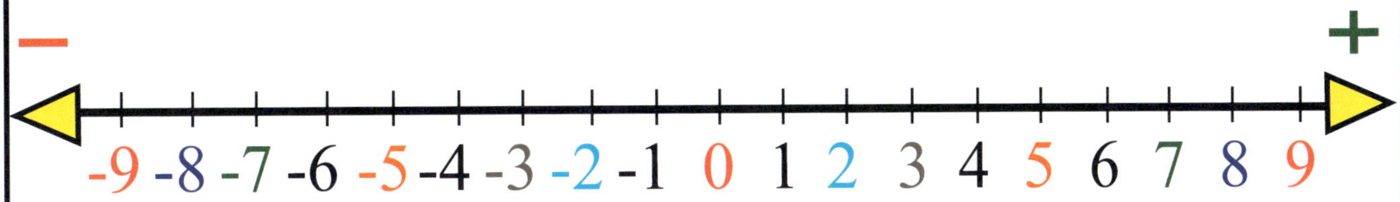

–
+
-9 -8 -7 -6 -5 -4 -3 -2 -1 0 1 2 3 4 5 6 7 8 9

Math Zambits arrived

and created a ring.

Would they demonstrate

this inequality thing?

> 2

The first example was

greater than two.

An open dot and a right arrow

was the thing to do.

> 2

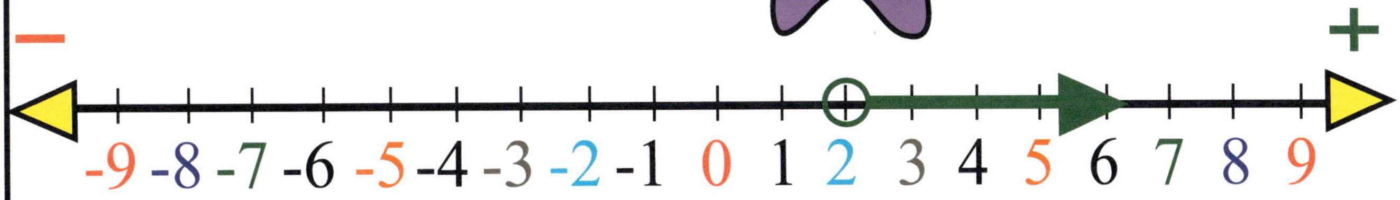

≥ 2

The second example was

greater than or equal to two.

A closed dot and a right arrow

was the thing to do.

≥ 2

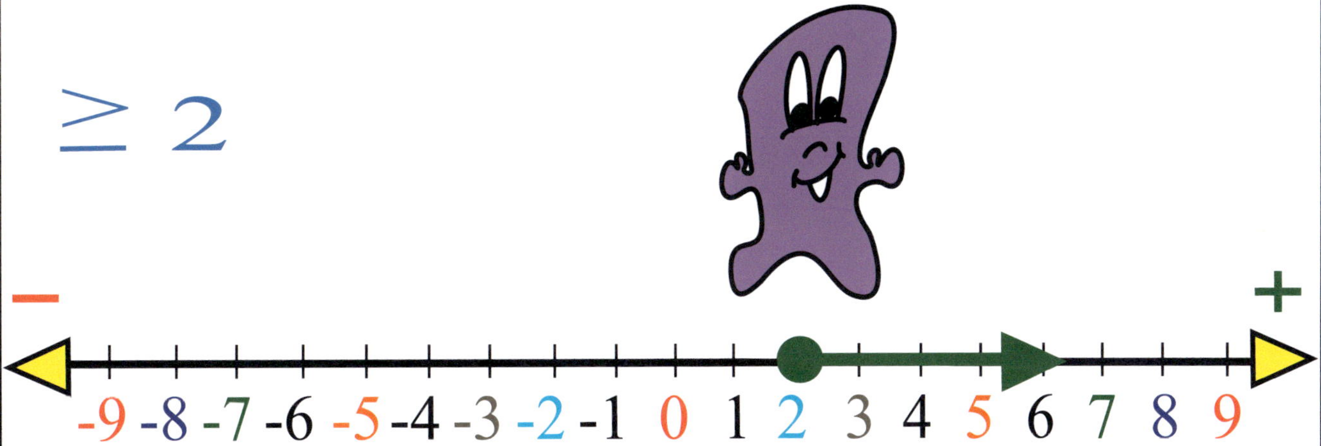

< -1

The next example was

less than negative one.

An open dot and a left arrow

is what was done.

< -1

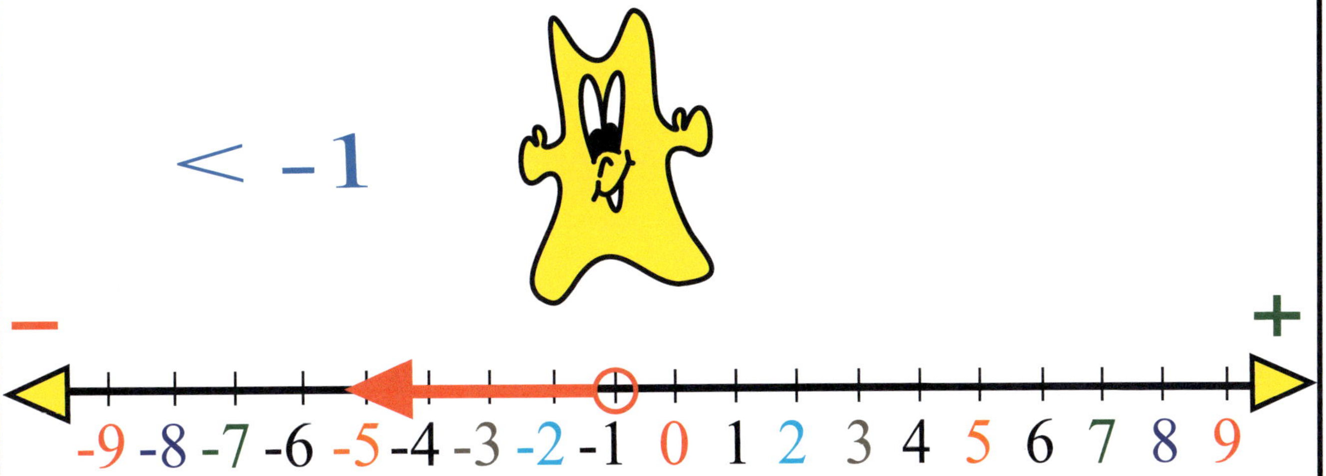

$\leq$ -1

The last example was

less than or equal to negative one.

A closed dot and a left arrow

is what was done.

≤ -1

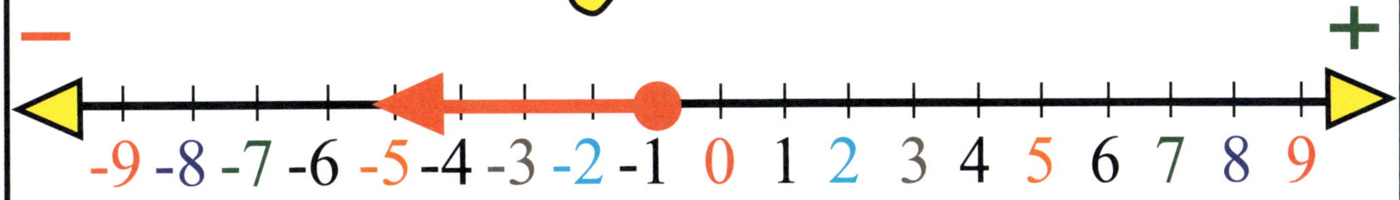

The Zambits had to go.

It was time to say goodbye.

I hope they will come back again

to give math another try.

I love to dream of Zambits.

That's exactly what I said.

I like to keep this inequality stuff

right inside my head.

THE END

≥ 4

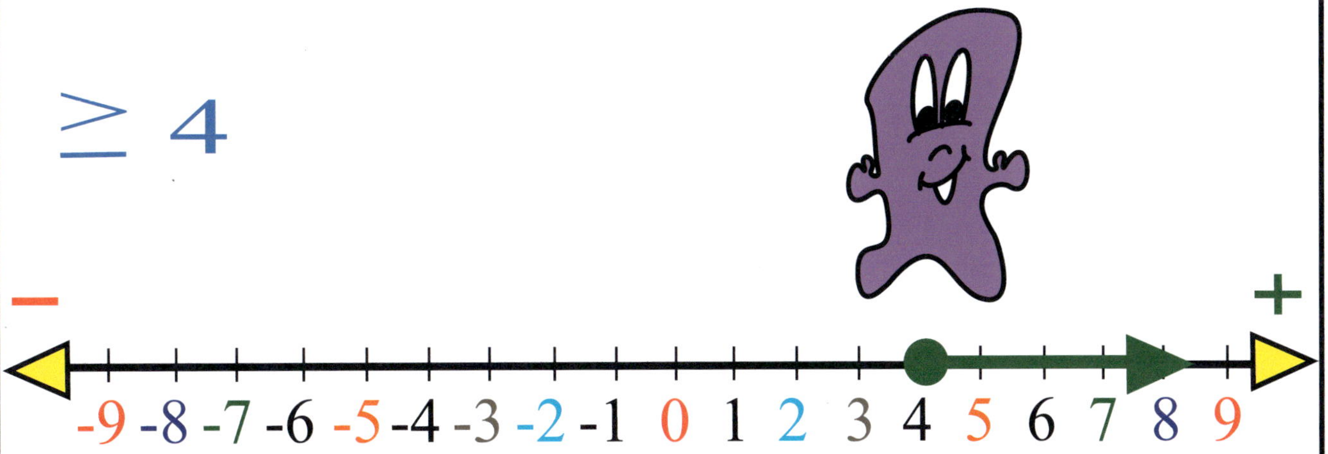

< -3

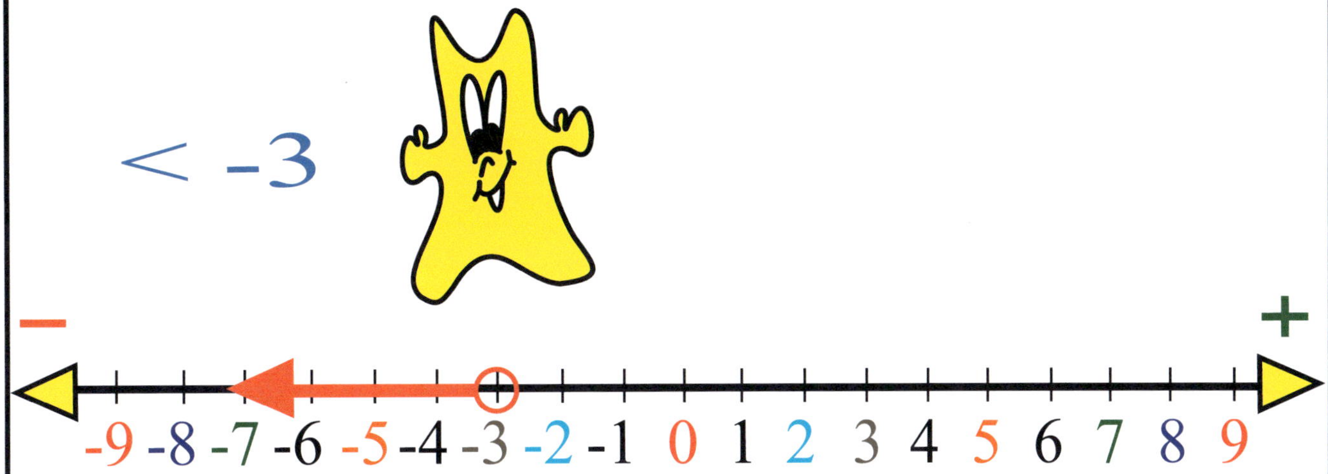

Made in the USA
Las Vegas, NV
07 November 2026

57650028R00021